Nuclear Power: How a Nuclear Power Plant Really Works!

by Amelia Frahm
illustrated by Andrew Handley

Nutcracker Publishing Company
www.NutcrackerPublishing.com

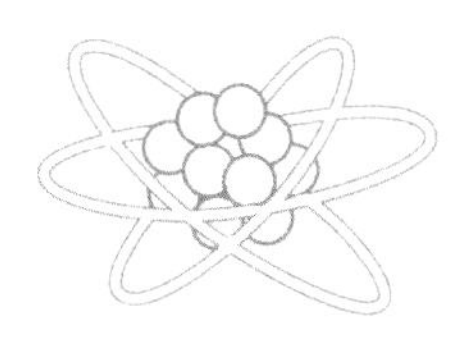
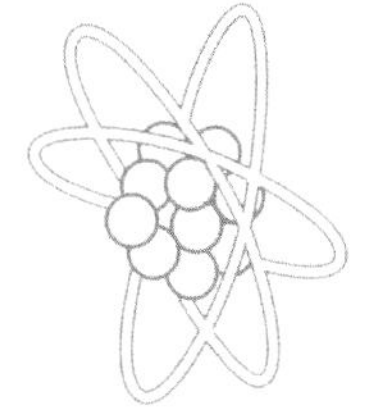
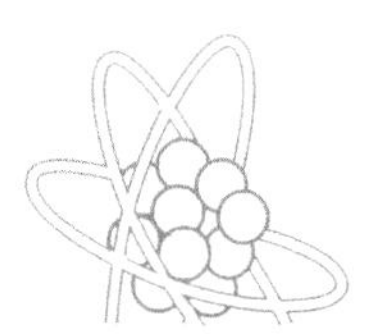

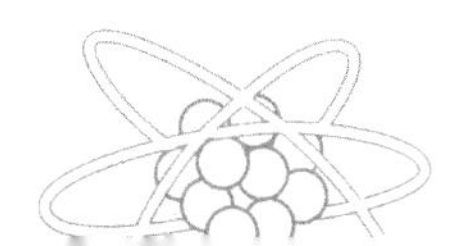

Nuclear Power: How a Nuclear Power Plant Really Works!

Published by: Nutcracker Publishing Company
Crack Open a Book!
www.NutcrackerPublishing.com
Info@NutcrackerPublishing.com

Special thanks to:

Edward Kee and other members of the American Nuclear Society Social Media Group

Kim Crawford
Progress Energy
Senior Communications Specialist
Harris Nuclear Plant

Tabitha Frahm
Marketing Assistant

Terry Godbey
Editor

Library of Congress Catalog Card Number: 2010933741
1. Juvenile/Fiction
2. Nuclear Power
3. Education/Science

ISBN 9780970575227

I wrote this for my Dimple Dumpling, who grew up before I could finish it. It's dedicated to my daddy who taught me to write, and my mother who taught me to read: Preston and Patricia Vaughan Solomon of Walnut Hill, Florida.

In memory — Laura Bouldin Karlman, 1961-2000.

At the Nukie Nuclear Power Plant, inside the nuclear reactor that sat beneath the big dome, a whole lot of fissioning was going on.

Where did it come from? Where did it go? Everyone who lived nearby was curious to know.

High atop the turbine building beside the dome, a chubby rat bragged to a pretty blue bird, “I know where fissioning comes from.”

“Well,” cawed the bird, who had flown in from nearby, “I know where electricity goes.” And they began to boast about who knew the most about Nukie Nuclear Power Plant.

It took them awhile to figure out they were being disagreeable over something they agreed about.

"The purpose of any power plant is to make electricity," chirped the bird from the transmission line where she was perched.

"I know that," declared the rat, "but a nuclear power plant makes electricity using fissioning. Who do you think causes it?" He gave his tail, which was missing its tip, a noisy tap.

"My tail is tip-less because Nukie Nuclear Power Plant was designed by a cat," wailed the rat.

The bird concurred. “I’m minus a few tail feathers of my own, and whose fault, do you think, is that?”

“Anything that goes wrong for a bird or rat, or raises questions or eyebrows, I blame on a cat,” declared the rat.

“I barely escaped my encounter with the female feline who caused the electricity-producing fission at Nukie Nuclear Power Plant,” the bird shrieked.

“It’s her fault my tail is tip-less,” the rat squeaked.

With that, the rat with a tip-less tail and the blue bird with a bald spot where her tail feathers should be decided there was one thing they both agreed upon: how a nuclear power plant really works.

Indeed, this particular fissioning chain reaction had powered the home of a tabby kitten.

On a day so hot that fish jumped out of the ocean already fried, most any cat with claws was down at the beach. Any cat except a red-furred one named Penelope, who did not wish to blister her whiskers.

Instead of going fishing at the beach, she went inside her house and turned on a fan and the air conditioner.

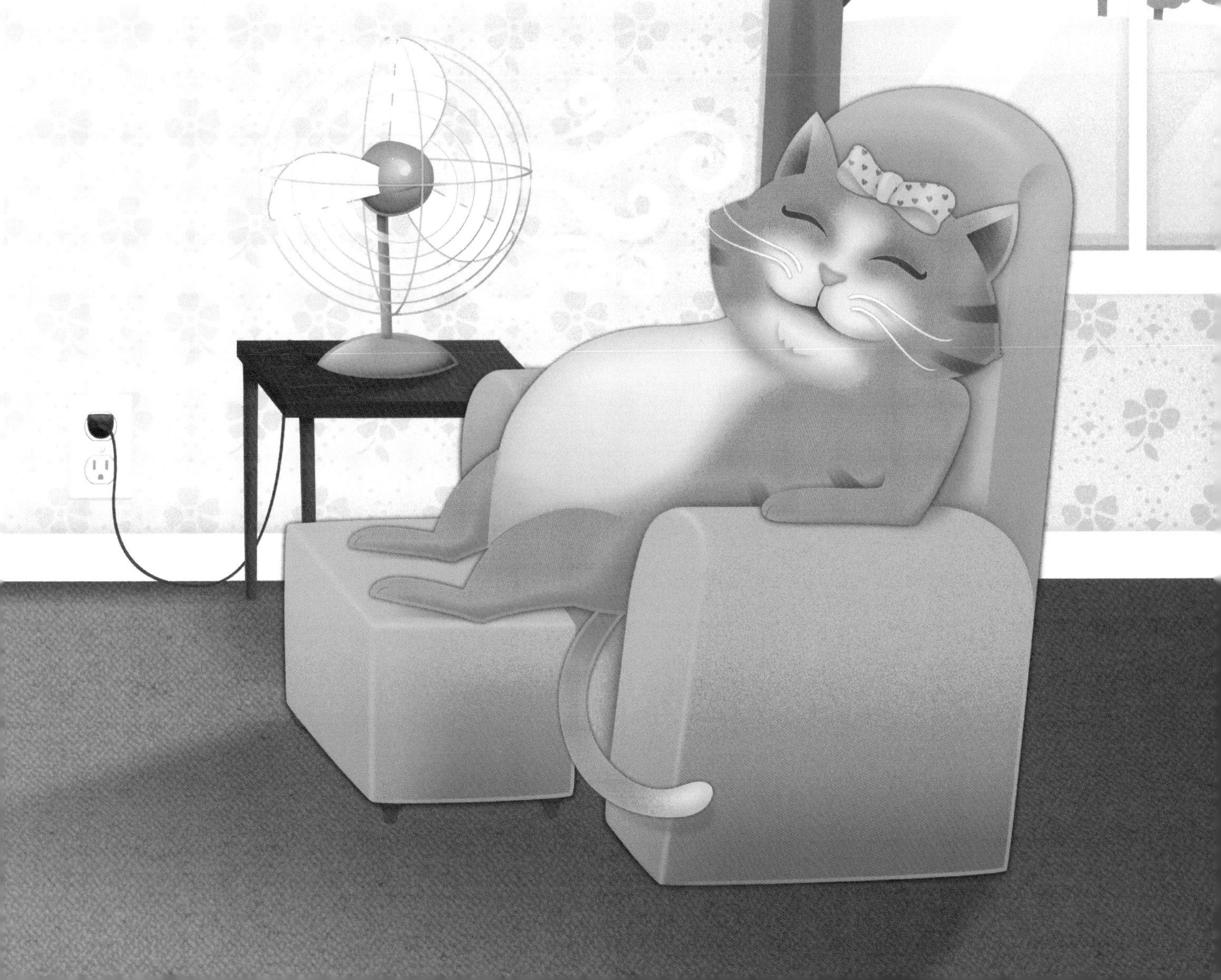

The pretty blue bird had flown over the kitty-crammed beach, keeping her feathers far out of reach. She noticed the absence of Penelope Cat and couldn't help wondering where she was at.

"I figured the only reason Penelope would not want to feast on fish was that she had in mind a scrumptious blue bird dish."

“It ruffled my feathers and roused my suspicions so I flew by Penelope’s house and peered into her kitchen,” the bird exclaimed as the rat listened.

Penelope was searching the contents of her refrigerator when she glanced out the window and glimpsed her investigator. Meowing wide, she waved her tail and motioned the bird inside.

"I didn't have time to hide! She turned on the stove the moment she eyed me," the bird cried.

"Resulting in a volt that caused a jolt inside the reactor at Nukie Nuclear Power Plant," alleged the rat. "A cat cannot cook without electricity, and Penelope gets hers from nuclear energy."

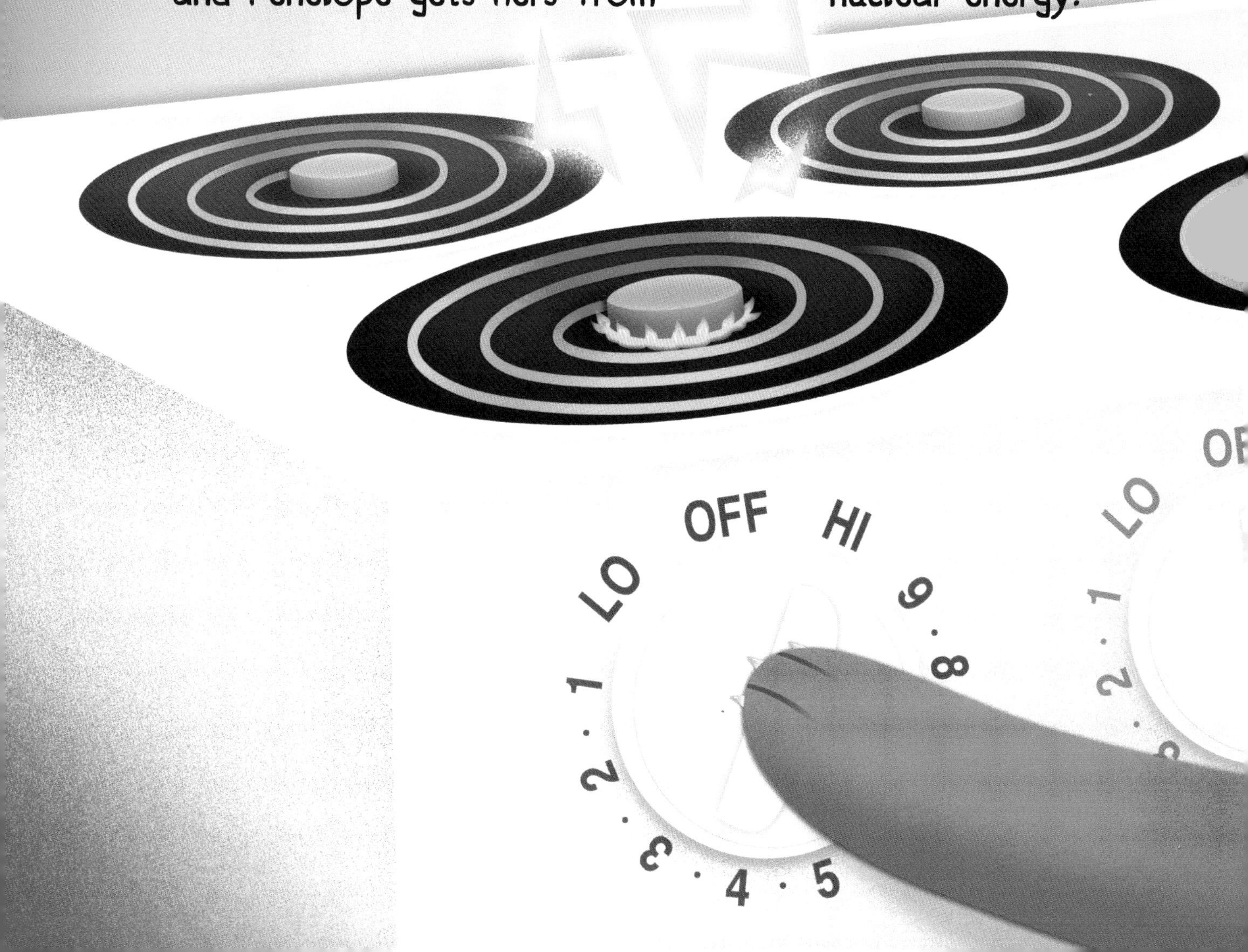

"I know the difference between a hot pot and a hot tub, and I do not bathe in either one," chit-chatted the bird. "The minute she invited me in I flew off so fast my tail struck the side of Penelope's house, and some of my pretty blue tail feathers got plucked out."

"Once I was safe above the ground, I took a look around at the town below. You may know where fissioning comes from, Rat, but I know where electricity goes."

The blue bird whispered, "It travels by transmission lines, quiet as a mouse, and into a meter box on the side of Penelope's house."

"How a nuclear power plant really works is just as I suspected," the rat interjected.

“The fissioning going on inside the reactor under the dome is used to produce electricity so Penelope Cat can laze around her home. That cat is basking under the air conditioner, cooking a gourmet dinner.”

“Meanwhile, some poor, starving rat has been tricked into demonstrating a fissioning chain-reaction and is dodging simulations of the uranium atoms that fuel the nuclear reactor.”

The rat paused to pant then continued to rant. "Atoms are so tiny and small you can't see them at all, but only a cat would symbolize uranium atoms using mousetraps loaded with neutrons made of orange Ping-Pong balls.

SNAP!

SNAP!

"Those neutrons looked like cheddar; I didn't think a little taste would matter, but it set off a mousetrap atom, and I ended up demonstrating a chain reaction. When a neutron hits an atom, the atom splits in two, and this split known as fission is what gave my tail its boo-boo."

"One atom fissions two,

two fission four,

and then there are more, more, and more," the rat moaned.

"A whole lot of tail-trimming fissioning is what goes on inside the reactor under the dome."

The bird ruffled her remaining tail feathers and twittered a groan.

"All that fission causes a lot of heat. It gets heated and it gets hot and the reactor vessel acts like a burner heating a great big pot. The water in the pot gets hotter and hotter," said the rat.

“—Until it screams with steam!” the blue bird hollered.

“The steam blows through pipes like a hurricane’s wind and causes the turbine building’s giant propellers to spin,” the rat shouted.

“Any leftover steam gets chilled by a blast of cold water from a cooling pond or cooling tower and then it’s recycled over and over,” the rat said.

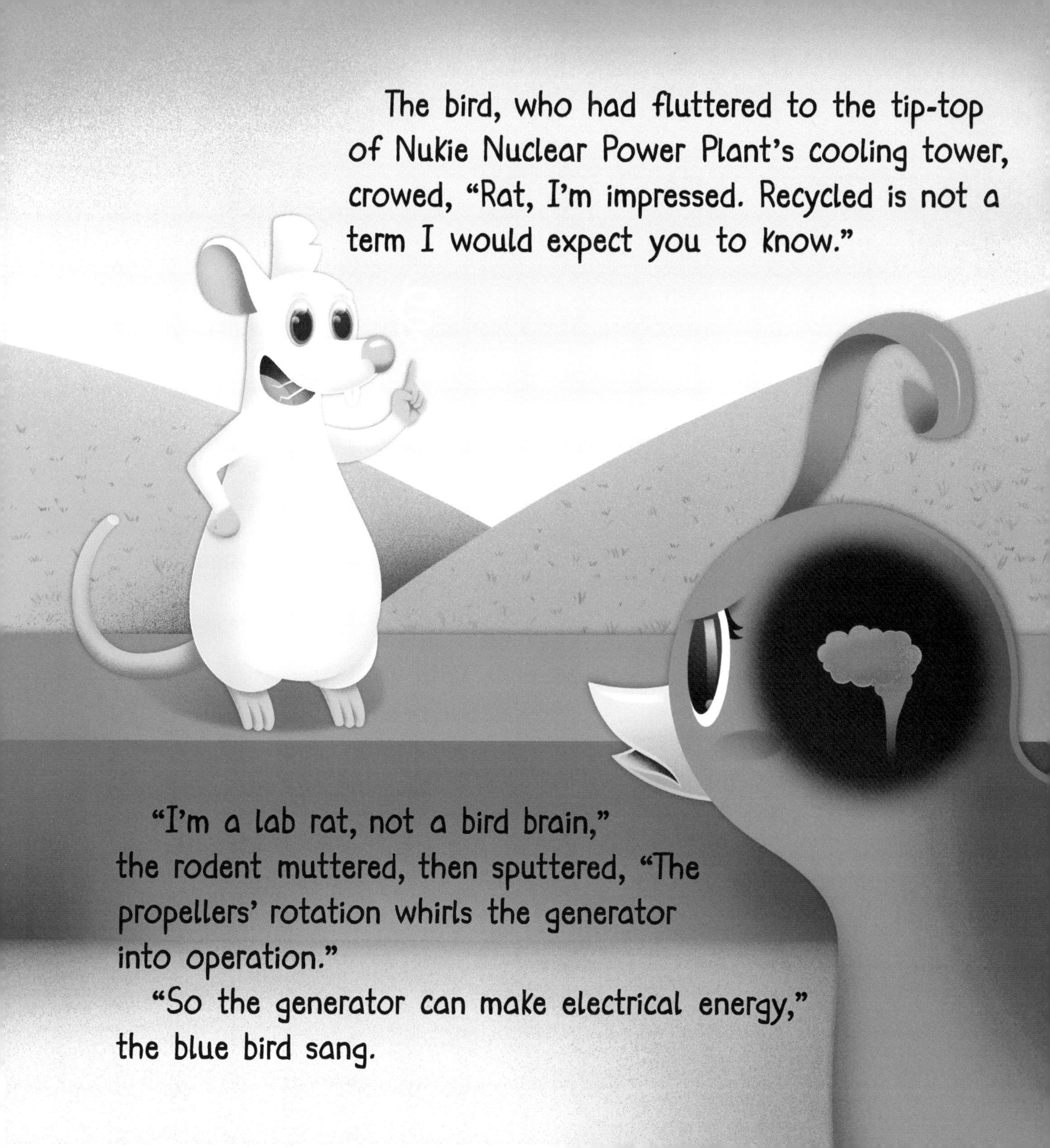

The bird, who had fluttered to the tip-top of Nukie Nuclear Power Plant's cooling tower, crowed, "Rat, I'm impressed. Recycled is not a term I would expect you to know."

"I'm a lab rat, not a bird brain," the rodent muttered, then sputtered, "The propellers' rotation whirls the generator into operation."

"So the generator can make electrical energy," the blue bird sang.

The rat gave his tip-less tail a bang. “Then it’s zapped on transmission wires and spewed to Penelope Cat, who uses electricity to bake a sweet blue bird pie or roast a tasty rat.”

And that, they believed, is how a nuclear power plant really works. As for Penelope Cat – unknown to our experts – she's a vegetarian. She likes to eat green beans, not bird or rat.

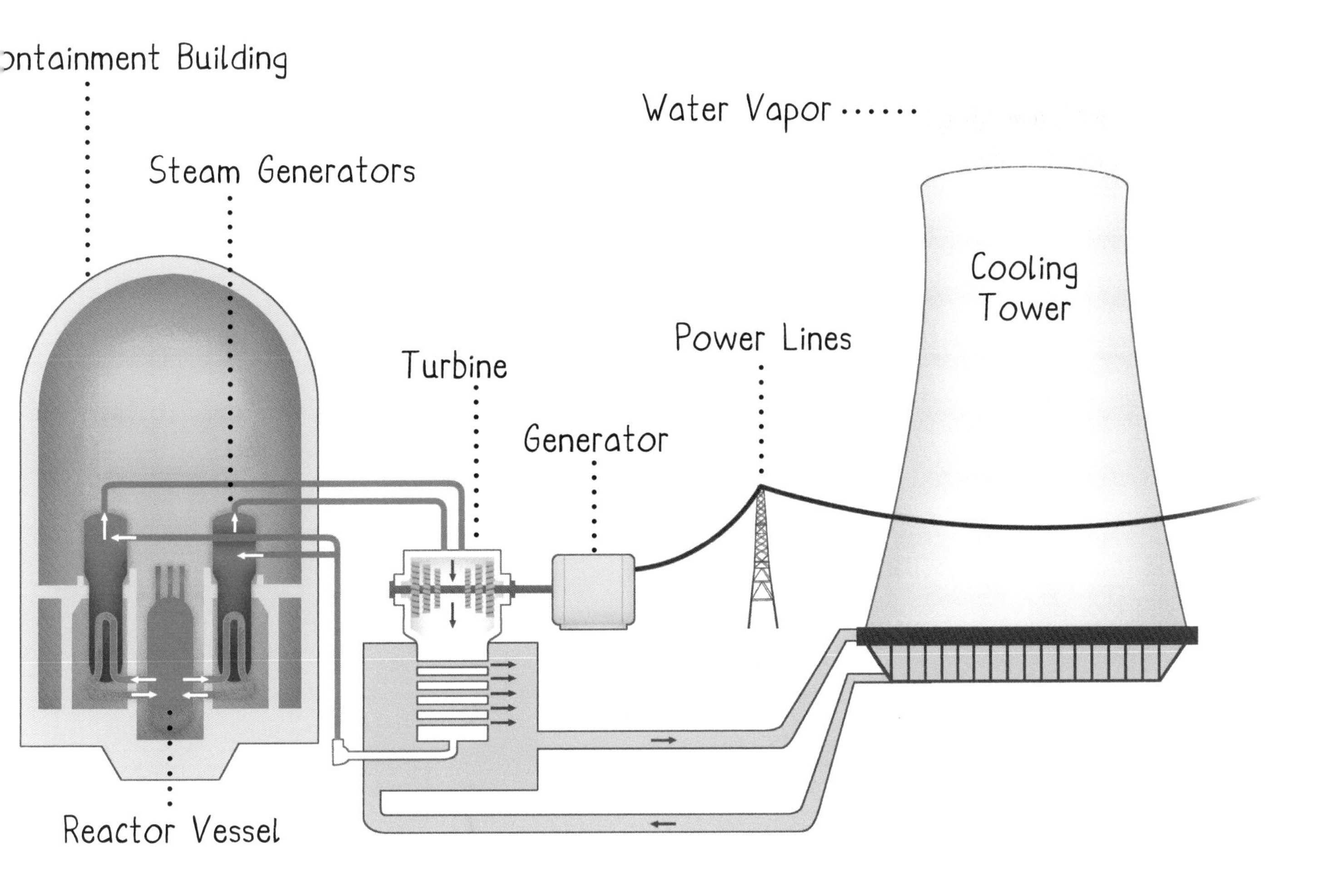

How a Nuclear Power Plant Works!

Glossary

Atom: A basic unit of matter, made up of neutrons and other particles.

Chain Reaction: A series of events in which the result of one event triggers another event, which in turn triggers yet another event.

Containment Building: A steel or concrete structure that houses a nuclear reactor. It is designed to keep everything safe in case of emergency.

Cooling Tower: A large cylindrical structure used to remove heat from water that has been used for cooling.

Fission: The act or process of splitting into parts.

Generator: A machine that converts mechanical energy into electricity.

Neutron: A particle found in the center of an atom; it has a neutral charge.

Nuclear Energy: Energy released by reactions within the nuclei of atoms.

Power Lines or Transmission Lines: Cables used to distribute electricity.

Reactor: A device to start and control a continual nuclear chain reaction.

Steam Generator: A machine used to convert water into steam.

Turbine: A rotary engine that takes energy from a fluid flow like steam, and converts it into useful work, or what is called mechanical energy.

Uranium: A metal found in most rocks. It is the fuel used inside a nuclear reactor.

For information on other books by Amelia Frahm, Educational Programs, and more, please visit

www.NutcrackerPublishing.com

Crack Open a Book!

CPSIA information can be obtained
at www.ICGtesting.com
Printed in the USA
238360LV00002B